Y0-BRO-025

Seasons in a Year

# SUMMER

AMY CULLIFORD

A Crabtree Roots Book

CRABTREE
Publishing Company
www.crabtreebooks.com

# School-to-Home Support for Caregivers and Teachers

This book helps children grow by letting them practice reading. Here are a few guiding questions to help the reader with building his or her comprehension skills. Possible answers appear here in red.

## Before Reading:

- What do I think this book is about?
  - *This book is about a season called summer.*
  - *This book is about things you can see in summer.*

- What do I want to learn about this topic?
  - *I want to learn what summer looks like.*
  - *I want to learn what people do in summer.*

## During Reading:

- I wonder why...
  - *I wonder why people like to fly kites.*
  - *I wonder why summer is so hot.*

- What have I learned so far?
  - *I have learned that people like to go to the pool in summer.*
  - *I have learned that people like to go for picnics in summer.*

## After Reading:

- What details did I learn about this topic?
  - *I have learned that there are many fun things to see in summer.*
  - *I have learned that summer is hot.*

- Read the book again and look for the vocabulary words.
  - *I see the word* ***kite*** *on page 4 and the word* ***picnic*** *on page 13. The other vocabulary words are found on page 14.*

What do you see
in **summer**?

I see a green **kite**.

I see a blue kite.

I see **sandals**.

I see a **pool**.

I see a **picnic**.

# Word List

## Sight Words

a
do
I
in
see
what
you

## Words to Know

**kite**

**picnic**

**pool**

**sandals**

**summer**

# 27 Words

What do you see
in **summer**?

I see a green **kite**.

I see a blue kite.

I see **sandals**.

I see a **pool**.

I see a **picnic**.

# Seasons in a Year

# SUMMER

Written by: Amy Culliford
Designed by: Rhea Wallace
Series Development: James Earley
Proofreader: Kathy Middleton
Educational Consultant: Christina Lemke M.Ed.
Photographs:
Shutterstock: sirtravelalot: cover; Elenamir: p. 1; Studio1One: p. 3, 14; RaduRazvan: p.4, 14; taptrofsnag: p. 9, 14; J.D.S.: p. 10-11, 14; NYS: ESBProfessional: p. 12

Library and Archives Canada Cataloguing in Publication
Title: Summer / Amy Culliford.
Names: Culliford, Amy, 1992- author.
Description: Series statement: Seasons in a year | "A Crabtree roots book".
Identifiers: Canadiana (print) 20200387022 |
Canadiana (ebook) 20200387049 |
ISBN 9781427134745 (hardcover) |
ISBN 9781427132710 (softcover) |
ISBN 9781427132758 (HTML) |
ISBN 9781427133137 (read-along ebook)
Subjects: LCSH: Summer—Juvenile literature.
Classification: LCC QB637.6 .C85 2021 | DDC j508.2—dc23

Library of Congress Cataloging-in-Publication Data
Names: Culliford, Amy, 1992- author.
Title: Summer / Amy Culliford.
Description: New York : Crabtree Publishing Company, 2021. | Series: Seasons in a year : a Crabtree roots book | Audience: Ages 4-6 | Audience: Grades K-1 | Summary: "Early readers are introduced to the summer season. Simple sentences and bright pictures feature summertime activities"-- Provided by publisher.
Identifiers: LCCN 2020049794 (print) |
LCCN 2020049795 (ebook) |
ISBN 9781427134745 (hardcover) |
ISBN 9781427132710 (paperback) |
ISBN 9781427132758 (ebook) |
ISBN 9781427133137 (epub)
Subjects: LCSH: Summer--Juvenile literature.
Classification: LCC QB637.6 .C85 2021 (print) | LCC QB637.6 (ebook) | DDC 508.2--dc23
LC record available at https://lccn.loc.gov/2020049794
LC ebook record available at https://lccn.loc.gov/2020049795

**Crabtree Publishing Company**
www.crabtreebooks.com 1-800-387-7650

Printed in the U.S.A./022021/CG20201130

 In Canada: We acknowledge the financial support of the Government of Canada through the Canada Book Fund for our publishing activities.

**Published in the United States**
**Crabtree Publishing**
347 Fifth Avenue, Suite 1402-145
New York, NY, 10016

**Published in Canada**
**Crabtree Publishing**
616 Welland Ave.
St. Catharines, Ontario L2M 5V6